NOTES TO USERS

THESE Notes are complementary to A.P. 2095 Pilot's Notes General, and assume a thorough knowledge of its contents. All pilots should be in possession of a copy of A.P. 2095 (see A.M.O. A718/48).

Additional copies may be obtained by the station publications officer by application on form 294A, in duplicate, to Command Head-quarters for onward transmission to A.P.F.S. (See A.P. 113). The number of the publication must be quoted in full—A.P. 2210C—P.N.

Comments and suggestions should be for-warded through the usual channels to the Air Ministry (T.F.2).

Air Ministry   A.P. 2210C—P.N.
March, 1949   *Pilot's Notes*
2nd edition
(Reprinted—January, 1951)
     Note.—This 2nd edition supersedes and cancels edition
          dated August, 1946.

METEOR 3

LIST OF CONTENTS

PART IV—EMERGENCIES

PART V—ILLUSTRATIONS

METEOR 3

PILOT'S CHECK LIST

(Excluding Checks of Operational Equipment.)

ITEM	CHECK
1. Weight and balance.	Ballasted if necessary.
2. Authorisation book.	Sign.
3. Form 700.	Sign.

External checks.

N.B.—Start at the port side of the nose and work clockwise around the aircraft.

ITEM	CHECK
4. Cockpit hood (port side).	Condition. Absence of cracks. Security.
5. External hood jettison handle.	Secure.
6. Nose wheel mechanical indicator.	Protruding.
7. Nose wheel.	Extension of oleo. Security of mudguard. Tyre for cuts and creep. Valve free. Condition of door and fairings.
8. External fire-extinguisher.	In position.
9. Cockpit hood (starboard side).	Condition. Absence of cracks. Security.
10. Starboard centre section.	Condition of leading edge.

ITEM	CHECK
11. Starboard nacelle.	All cowlings secure. Intake cover removed.
12. Starboard under-carriage.	Condition of doors. Brake leads secure. Tyre for cuts and creep. Valve free. Chock in position.
13. Starboard mainplane.	Condition of leading edge.
14. Starboard navigation light.	Condition.
15. Starboard aileron.	Condition. External control lock removed.
16. Starboard mainplane.	Condition of upper and lower surfaces. Picketing ring removed. Condition of flaps and air brakes.
17. Starboard nacelle.	Jet pipe cover removed.
18. Starboard fuselage.	Condition.
19. External aerial.	Secure.
20. Fin.	Condition of leading edge.
21. Starboard tailplane.	Condition. Leading edge.
22. Starboard elevator.	Condition. Trimmer. External control lock removed.

6

ITEM	CHECK		ITEM	CHECK
23. Rudder.	Condition. Trimmer. External control lock removed.	37.	Port centre section.	Condition of leading edge.
24. Tail light.	Condition.	38.	Ventral drop tank.	Secure.
25. Port elevator.	Condition. Trimmer. External control lock removed.	39.	Dispersal area.	All clear around aircraft.
26. Port tailplane.	Condition. Leading edge.		**Internal checks.**	
		40.	Internal control locks.	Removed.
27. Emergency skid.	Condition.			
28. Port fuselage.	Condition.	41.	Undercarriage selector lever.	Down.
29. Port nacelle.	Jet pipe cover removed.	42.	Pilot's seat.	Adjusted for height.
30. Port mainplane.	Condition of upper and lower surfaces. Picketing ring removed. Condition of flaps and air brakes.	43.	Rudder pedals.	Adjusted for length.
		44.	Flying controls.	Full and correct movement.
			Cockpit checks.	
31. Port aileron.	Condition. External control lock removed.		N.B.—Work from left to right	
		45.	Port H.P. cock.	On.
32. Port navigation light.	Condition.	46.	Port L.P. cock.	On.
33. Pressure head.	Cover removed.	47.	Balance cock.	Shut.
34. Port mainplane.	Condition of leading edge.	48.	Crowbar.	In position.
35. Port nacelle.	All cowlings secure. Intake cover removed.	49.	Air pressure gauge.	Supply. Delivery to each wheel brake.
36. Port undercarriage.	Ground/flight switch to flight. Condition of doors. Brake leads secure. Tyre for cuts and creep. Valve free. Chock in position.	50.	Rudder trim control.	Full and correct movement.
		51.	Elevator trim control.	Full and correct movement.
		52.	Air brakes lever.	Off (forward).
		53.	Ventilation control.	Cold.

	ITEM	CHECK			ITEM	CHECK
54.	Port low pressure pump switch.	Off.		70.	Pressure-head heater switch.	Off.
55.	Starboard low pressure pump switch.	Off.		71.	Identification lights switch.	As required.
56.	Fuel pressure warning lights.	On.		72.	R.I. compass switch.	Off.
57.	Landing lamp switch.	Operation. Retract lamp.		73.	Hood.	Operation of winding handle.
58.	Flap lever.	Neutral.		74.	Hydraulic handpump.	Exhaust accumulator pressure. Pump flaps down and up.
59.	Under-carriage indicator.	Operation. Green lights on.		75.	Wind-screen de-icer.	Operation.
60.	Under-carriage warning light.	Out.		76.	Starboard H.P. cock.	On.
61.	Flap indicator.	Compare with position of flaps.		77.	Starboard L.P. cock.	On.
62.	Ventral tank transfer control.	Off. Handle in.		78.	Pilot's harness.	Adjust. Test lock.
63.	Altimeter.	Set.		79.	Ground/ flight switch.	Ground.
64.	Direction indicator.	Caged.			**Start the engines** (see para. 36).	
65.	Hood jettison handle.	In.		80.	Generator failure warning light.	Out.
66.	Fuel gauges.	Contents.		81.	Fuel pressure warning lights.	Out.
67.	Oxygen.	Delivery.		82.	Vacuum change-over cock.	Operation.
68.	Generator failure warning light.	On.		83.	Flaps.	Lower. Indicator reading. Raise. Selector neutral.
69.	Navigation lights switch.	As required.				

ITEM	CHECK
84. R.I. compass.	Switch on.
85. Direction indicator.	Set with R.I. compass. Uncage.
86. Radio.	Test V.H.F. and other radio aids Check altimeter setting with control.

Checks before and during taxying.

ITEM	CHECK
87. Chocks.	Clear.
88. Taxying.	As soon as possible test brakes. Direction indicator for accuracy. Artificial horizon for accuracy. R.I. compass for accuracy against known heading. Temperatures and pressures. Pressure head-heater on if necessary.

Checks before take-off.

ITEM	CHECK
89. Trim : Elevator	Neutral to ½ div. nose up.
Rudder.	Neutral.
90. Fuel.	Check contents. H.P., L.P., cocks on. Low pressure pumps on. Ventral tank transfer control off. Balance cock shut.
91. Flaps.	¼ down. Selector neutral.
92. Pneumatic supply.	At least 200 lb./sq. in.

ITEM	CHECK
93. Air brakes.	Closed (off).
94. Sliding hood.	Closed.
95. Direction indicator	Set to R.I. compass and uncaged.
96. Harness.	Adjusted and locked.

Checks in flight as necessary.

Checks before landing.

When entering the circuit :—

ITEM	CHECK
97. Pneumatic supply pressure.	More than 120 lb./sq. in. Check pressure to brakes.
98. Fuel.	Check contents. Balance cock open if necessary.

Reduce speed to 175 knots.

ITEM	CHECK
99. Harness.	Locked.
100. Under-carriage.	Down. Green lights on. Nosewheel mechanical indicator showing.
101. Air brakes.	As required.
102. Flaps.	As required.

Checks after landing.

When clear of the landing area :—

ITEM	CHECK
103. Pneumatic supply.	Sufficient pressure for taxying.
104. Flaps.	Up. Selector neutral.
105. Air brakes.	Closed (off).
106. Pressure head-heater.	Off if necessary.

ITEM	CHECK	ITEM	CHECK
On reaching dispersal.		114. Brakes.	Off.
107. H.P. cocks.	Off.	115. Internal control locks.	On.
108. Low pressure pumps.	Off.		
When the engines have stopped.		**After leaving the aircraft.**	
109. L.P. cocks.	Off.	116. Ground/ flight switch.	Ground.
110. Electrical services.	All off.		
111. Radio.	Off.	117. Pressure head.	Cover on.
112. Direction indicator.	Caged.	118. Form 700.	Sign if necessary.
113. Chocks.	In position.	119. Authorisation book.	Sign.

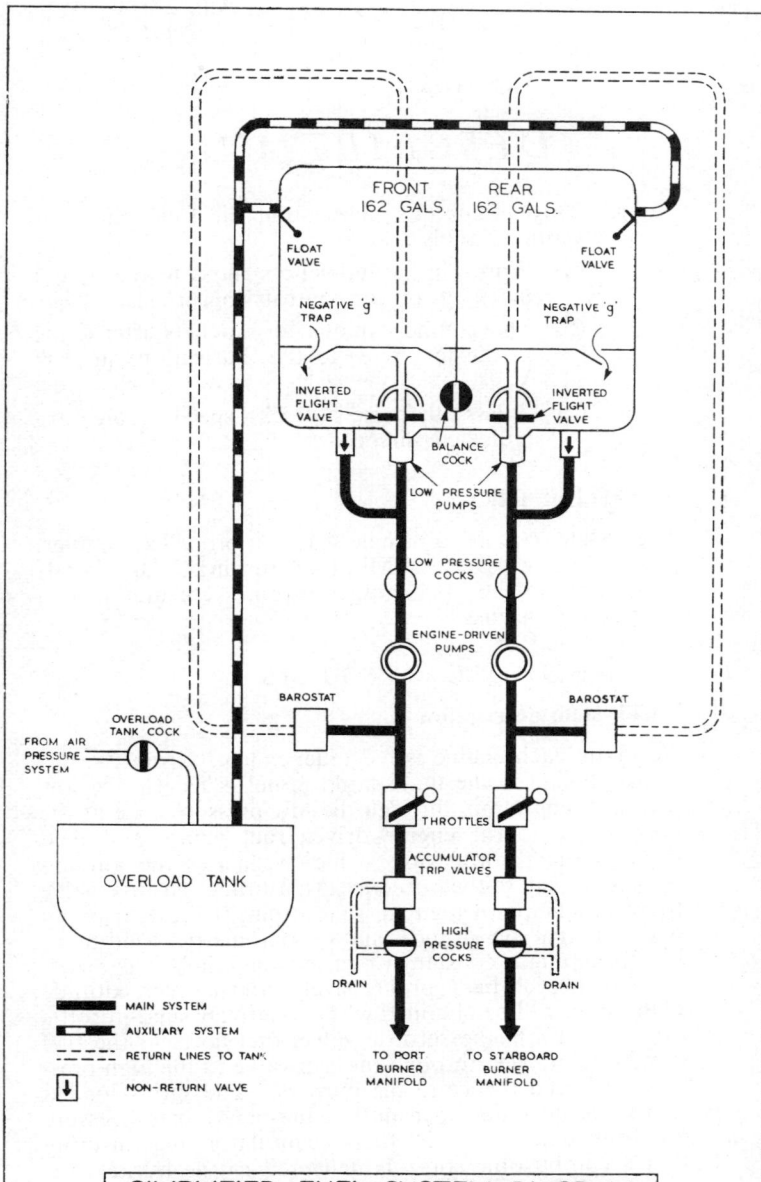

SIMPLIFIED FUEL SYSTEM DIAGRAM

PART I

DESCRIPTIVE

NOTE.—Throughout this publication the following conventions apply :—

   (a) Words in capital letters indicate the actual markings on the controls concerned.

   (b) The numbers quoted in brackets after items in the text refer to the illustrations in Part V.

   (c) Unless otherwise stated all speeds quoted are indicated airspeeds.

INTRODUCTION

The Meteor 3 is a single-seat jet-propelled fighter, powered by two Derwent Mk. 1 gas turbines. On aircraft subsequent to EE. 465 the gas turbines are mounted in long chord nacelles.

FUEL AND OIL SYSTEMS

1. Fuel system description

Normally each engine is fed independently from its own compartment of the main tank. Fuel is fed by the low pressure fuel pump, through the low-pressure cock at the tank outlet to the engine driven fuel pump, and then through the throttle valve which regulates the amount of fuel passing to the burners. An altitude sensitive valve (barostat) helps to regulate the output of the fuel pump with changes of altitude and/or barometric conditions, but to maintain constant r.p.m. on the climb it is necessary to throttle back progressively, and at high altitude high r.p.m. will be obtained with a relatively small throttle opening. From the throttle valve, fuel flows to the fuel accumulator, then through the trip valve to the high pressure cock and thence to the burners. The spring-loaded trip valve does not open until a pre-determined pressure has been built up in the fuel accumulator thus ensuring that a combustible spray is delivered to the burners for the initial start. The trip valve, and high pressure cock which incorporates a by-pass back to the pump inlet as well as a drain to atmosphere, are located in the body of the fuel accumulator. When the high pressure cock is closed the by-pass to the pump inlet and the drain to atmosphere are opened ; in addition a spring-loaded valve in the lowest combustion chamber opens when the pressure therein falls to a low value, allowing any surplus fuel in the lower combustion chambers, which are inter-connected, to drain to atmosphere.

2. **Fuel tanks**

(i) The main fuel tank, which is self-sealing, holds 325 gallons and is divided into two equal compartments. The front compartment normally feeds No. 1 engine and the rear compartment No. 2 engine, but they may be inter-connected by a balance cock which is under the control of the pilot. Each compartment contains an " inverted-flight " trap and valve which under negative " g " conditions ensure a supply of fuel to the engines for about 15 seconds.

(ii) On early aircraft a fixed 100-gallon ventral tank may be fitted and the fuel from this tank is transferred to the main tank by air pressure from the exhaust side of the vacuum pumps. On later aircraft this tank is replaced by a 180-gallon ventral drop tank, the fuel from which is transferred to the main tank by air pressure from the engines.

Float valves in each compartment of the main tank cut off the supply of fuel from the ventral tank when the main tank is nearly full.

3. **Fuel cocks**

(i) The low (1) and (47), and the high-pressure, (2) and (46), cocks for each engine, are on either side of the pilot's seat, and are marked L.P. and H.P. The outer lever of each pair controls the H.P. cock.

The L.P. cocks cut off the flow of fuel from the main tank ; they should not be closed to stop the engines, as this action will starve the engine-driven fuel pumps and fill the pipe-lines with air. The appropriate L.P. cock should, however, be closed in the event of an engine failure (see para. 53). The H.P. cocks cut off the supply to the engine burners and should be used for stopping the engines, and also in the event of engine failure.

(ii) *Ventral tank transfer control:* The transfer control which permits air to feed into the ventral tank is operated by a handle on the top left-hand side of the front panel. On early aircraft which have a 100-gallon ventral fixed tank, the handle is pulled out to transfer fuel to the main tank. On later aircraft which have a 180-gallon ventral drop tank, the handle has a dual function: it is rotated anti-clockwise to transfer fuel to the main tank and is pulled out to jettison the drop tank. It cannot be pulled out unless it is in the OFF (vertical) position.

On all aircraft, when the handle is set to transfer fuel from the ventral tank, a red warning light comes on until the pressure in the transfer pipeline rises above 2 lb./sq. in; it will then go out.

When all the fuel has been transferred from the ventral tank the warning light will come on and remain on. After it has remained on for at least 5 minutes, the control handle should be returned to the OFF position, when the warning light will go out again.

(iii) *The balance cock:* The balance cock (20) on the cockpit floor just aft of the rudder trimming tab control, is pulled up to interconnect the two compartments of the main tank. If an engine should fail, opening the balance cock will enable both compartments to feed the live engine. The H.P. and L.P. cocks of the failed engine should both be turned OFF before opening the balance cock.

4. Low pressure pumps

The low pressure pump switches (12) are on the engine starting panel below the left-hand side of the front panel. On early aircraft these switches are labelled SUPER-CHARGER SAFETY SWITCHES; on later aircraft they are labelled START AND TANK PUMP.

5. Fuel pressure warning lights

Two fuel pressure warning lights are fitted to the left of the flap lever. The appropriate light comes on when the fuel pressure from the low-pressure pump falls appreciably below normal. Should a light come on during flight, indicating pump failure, it will be impossible to obtain maximum r.p.m. at high altitudes on the engine concerned.

6. Fuel contents gauges

Two electrical fuel contents gauges (28), one for each main tank compartment, are fitted on the right-hand side of the front panel, and indicate whenever electrical power is available.

7. Oil system

An oil tank is fitted in each engine nacelle. Each tank holds 22 pints of oil with 7 pints air space.

ENGINE CONTROLS

8. Throttle controls

The throttle levers (5) are on the cockpit left-hand wall. No friction damping control is fitted.

9. Engine starting system

The starting cycle is controlled automatically by time switches. The shielded pushbuttons (13) on the left-hand switch panel, when pressed for about two seconds and then released, operate the time switches bringing into operation the starter motors and the igniter plugs which are used only for starting combustion ; current to the starter panel is automatically cut off after 30 seconds. The throttle must be closed and the low-pressure pump switch must be ON before the starting circuit is completed.

10. Relighting switches

The two pushbuttons (10) marked BOOST TEST or, on later aircraft, ENGINE RELIGHT SWITCH, at the top of the left-hand switch panel should be used to energise the igniter plugs for relighting the engines in flight ; the normal starting system should never be used.

11. Engine instruments

The following engine instruments are provided :—
R.p.m. indicators (34),
Jet pipe temperatures gauge (31),
Burner pressure gauges (32),
Oil temperature and pressure gauges (33).

NOTE.—The scale on the burner pressure gauge is such that accurate reading is difficult. Furthermore the burner pressure is influenced by height and r.p.m. ; hence it is not possible to lay down definite readings for given conditions of flight.

MAIN SERVICES

12. Hydraulic system

A hydraulic pump, driven by No. 2 engine, operates the :—
Undercarriage.
Flaps.
Air brakes.

A hydraulic accumulator is fitted and provides a reserve of pressure for operating any of the services when the engine is not running ; it is designed to provide sufficient pressure for the operations described in para. 54 (iv).
A handpump (30) on the right of the seat will operate all services through the normal pipelines, and may be used in the event of failure of the engine-driven pump and lack of accumulator pressure.

13. Pneumatic system

Two storage cylinders are charged before flight to 450 lb./sq. in. for operating the brakes and cocking the guns. On early aircraft the available pressure is shown on the gauge fitted just below the cockpit ventilation control, while the pressure at each wheel is shown on a triple pressure gauge (4) on the cockpit left-hand wall level with the pilot's shoulder. On later aircraft the nose-wheel brake is not fitted and the available pressure is shown by the top needle of the triple pressure gauge. There is no engine-driven compressor.

PART 1 — DESCRIPTIVE

14. Electrical system

(i) A 24-volt generator on No. 1 engine charges two batteries which in turn supply the whole of the electrical system.

(ii) The ground starter battery socket and ground/flight switch are in No. 1 engine nacelle.

(iii) A generator failure warning light, which comes on when the generator is not delivering current to the accumulators owing to a fault, is, when fitted, mounted on the switch panel on the cockpit right-hand wall.

15. Vacuum system

A vacuum pump is fitted on each engine and the selector cock (48) is mounted on the cockpit right-hand wall below the fuel cocks.

AIRCRAFT CONTROLS

16. Flying controls

The flying controls are conventional and the rudder pedals can be adjusted for reach by pulling out (on later aircraft, pushing in) the knob (35) on the left-hand side of the front panel. On early aircraft the pedals are not interconnected and care must be taken to ensure that they are adjusted equally.

17. Flying controls locking gear

The flying controls are locked in the neutral position by four rods fitted at each end with small pegs.
Two of the rods are used to lock the rudder pedals to the elevator torque tube and the second two to connect the control column to attachment points on the cockpit right-hand wall and the bulkhead behind the pilot's seat. When not in use the locking gear is stowed on the decking beneath the fixed part of the cockpit hood.

18. Trimming tabs

The elevator trimming tabs are controlled by a handwheel (18) on the left-hand side of the cockpit. The rudder trimming tab is controlled by a smaller handwheel (19) aft and to the left of the elevator trimming tab con-

trol. Both controls work in the natural sense and each
has an adjacent indicator (17).

19. Undercarriage control

The undercarriage selector lever (9) on the left-hand side
of the front panel has two positions, UP and DOWN.
It cannot be set to UP while the weight of the aircraft
is on the wheels, and there is no emergency override
switch to permit the undercarriage to be raised on the
ground.

20. Undercarriage position indicators

(i) A standard electrical visual indicator (21) on the left-
hand side of the front panel operates as follows :—

Undercarriage locked down ...	...	3 green lights
Undercarriage between locks	...	3 red lights
Undercarriage locked up	...	All lights out

(ii) A warning light on the front panel comes on if any wheel
is not locked down and either throttle is less than one-
third open. At moderate and high altitudes this light will
probably remain on continuously since the throttles are
usually less than one-third open in cruising flight.

(iii) *Nosewheel mechanical indicator*.   When the nosewheel
is down a small rod protrudes through the nose of the
aircraft.

21. Flaps control and indicator

(i) The flaps selector lever (8), on the left-hand side of the
front panel, has three positions—UP, NEUTRAL,
DOWN. To obtain an intermediate position of the flaps,
the lever should be moved to DOWN, and returned to
NEUTRAL when the flaps have reached the desired
setting. The selector lever should always be returned to
NEUTRAL after each operation is complete.

(ii) The flaps position indicator (22) is on the front panel
above the undercarriage visual indicator.

22. Air brakes

The air brakes control (7) is on the cockpit left-hand
wall below the throttle levers ; the control is pulled back
to open the brake flaps.

23. Wheel brakes

The control lever (36) and parking catch are fitted on the control column. Differential braking is obtained by use of the rudder pedals when the control lever is operated. On some early aircraft a nose wheel brake is fitted.

OPERATIONAL CONTROLS

24. Gyro gun sight

The selector and master switches (42) for the gyro gun sight (25) are mounted on the switch panel on the cockpit right-hand wall, while the ranging control is incorporated in the top of the No. 1 engine throttle lever.

25. Gun firing

The guns are fired electrically by the wobble button (37) on the control column. When the safety flap is open, pressing any part of the button fires all four guns, together with the cine-camera if the camera master switch is ON. On some aircraft a rounds counter (14) is mounted on the cockpit left-hand wall.

26. Cine-camera

The master switch is on the switch panel on the cockpit right-hand wall. The cine-camera operates with the guns, or can be fired independently by the cine-camera button which is visible when the gun safety flap is at SAFE. The footage indicator is on the cockpit left-hand wall.

OTHER CONTROLS

27. Cockpit ventilation and oxygen system

(i) On early aircraft the control lever on the cockpit left-hand wall above the throttle levers has three positions marked COLD, HOT and PRESSURE. The cockpit is not pressurised and there is no heating system ; the COLD position gives maximum flow of cold ventilating air, the HOT position gives a reduced flow of cold air while the PRESSURE position shuts off all ventilation from the cockpit. On later aircraft the control lever described above has two positions only, **HOT and COLD**.

(ii) The pilot's flexible oxygen pipe is clipped to the side of the seat and a standard Mk. 11 series regulator (29) is on the front panel.

28. Sliding hood

The sliding hood is opened and closed by the crank handle (43) on the cockpit right-hand wall. The hood is locked in the closed position, the open, or any intermediate position, by folding the crank outwards and engaging it in the slot below the right-hand side panel of the windscreen. Provision is made for jettisoning the hood (see para. 58).

29. Windscreen

The windscreen is of the " dry air sandwich " type. It should remain free from internal misting while the silica-gel remains active. The silica-gel containers should be replaced when the crystals turn from blue to pink.

30. Windscreen de-icing pump

The windscreen de-icing pump (49) and flow control for the de-icing fluid are on the cockpit floor to the right of the pilot's seat.

31. R.I. compass

The switch for the R.I. compass indicator (23) is the rearmost of those shown at (40) on the right of the cockpit wall.
There is no " stand-by " magnetic compass.

32. Cockpit lighting

Red floodlights. The three instrument panel lights are controlled by a dimmer switch on the top right-hand side of the front panel. Two auxiliary lights, one to illummate the trimmer handwheels and the other the compass, are controlled by a dimmer switch on the under surface

FINAL CHECKS FOR TAKE-OFF

TRIM	... ELEVATOR : NEUTRAL TO $\frac{1}{2}$ DIV. NOSE UP
	RUDDER : NEUTRAL
FUEL	... L.P. COCKS ON
	H.P. COCKS ON
	L.P. PUMPS ON
FLAPS	... UP OR $\frac{1}{3}$ DOWN
AIR BRAKES	... CLOSED

FINAL CHECKS FOR LANDING

BRAKES   ... OFF
                         CHECK PRESSURES

WHEELS   ... LOCKED DOWN

FLAPS   ... FULLY DOWN
                               ON FINAL

of the left-hand coaming.

Ultra-violet lights. Two u.v. flood lights, one on each side of the cockpit, illuminate the fluorescent instruments. Both are controlled by a dimmer switch on the left-hand side of the front panel.

Emergency light. An emergency light is mounted above the centre instrument panel. Its switch (45) and accumulator are on the right-hand side of the cockpit.

33. Seat-raising mechanism and harness release

The seat can be adjusted for height by the lever (50) on its right-hand side. The harness is released by a lever (44) on the right-hand side of the cockpit. After the lever has been returned to its normal position, the harness returns automatically to the locked position when the pilot leans back.

34. Footsteps

Footsteps and handholds are provided on the port side of the front fuselage to give access to the cockpit. A retractable footstep is lowered by pulling down the external handle and is automatically retracted when the undercarriage selector lever is moved to UP; no attempt should be made to push the step into its recess by hand. A second footstep which can be used as a first handhold, has a spring-loaded flap and is midway up the fuselage below the hood. The top handhold, which also has a spring-loaded flap, is in the top of the fuselage directly above the retractable footstep.

PART II
HANDLING

35. Management of the fuel system

The low pressure pumps must be ON for starting and should remain ON whenever the engines are running. When a ventral tank is carried its contents should be transferred to the main tank early in flight. If the two compartments of the main tank have been used or replenished unevenly, the balance cock should be opened.

36. Starting the engines

(i) Before starting the engines carry out the external, internal and cockpit checks detailed in the Pilot's Check List.

(ii) Have a ground starter battery plugged in and have the ground/flight switch set to GROUND.

(iii) Ensure that no personnel or loose equipment are in the vicinity of either the air intakes or the wake of the jet pipes. The engines should not be started with the tail into a strong wind.

(iv) Ensure :—
Throttle levers ... ... fully back
H.P. and L.P. cocks ... ... ON

(v) Start No. 1 engine first as it drives the generator. Switch on the low pressure pump, then when the fuel pressure warning light goes out, press the starter pushbutton for about 2 seconds and release it.

(vi) The engine should accelerate to 5,000 to 6,000 r.p.m. without further throttle adjustment. The jet pipe temperature may momentarily exceed the idling limit, but it should soon settle down to not more than 500°C. Do not open the throttle before idling speed is attained.

(vii) Repeat the above procedure for No. 2 engine and when both engines are running at idling r.p.m. and not before, have the ground/flight switch set to FLIGHT and the ground starter battery disconnected.

NOTE.—(a) It is important that the ground starter battery is fully charged and switched on the whole time during the starting sequence, as the engine requires assistance from the starter to accelerate up to idling r.p.m.

(b) Although it is possible to start the engines on the aircraft batteries, this should not be done except in emergency.

(c) If the engine accelerates very slowly up to idling r.p.m. and there is jet pipe resonance, this may be cured by partially closing the H.P. cock and immediately opening it when the resonance ceases. In no circumstances should resonance be allowed to continue unchecked. If resonance has necessitated shutting down the engine, sufficient time must be allowed for excess fuel to drain away before a restart is attempted.

(d) Should the jet pipe temperature reach 600°C., close the H.P. cock to stop the engine.

(e) The ground battery should not be disconnected until the ground/flight switch has been set to FLIGHT.

(viii) In the event of a " wet start," i.e., opening of the trip valve without light-up occurring, proceed as follows :—

(a) Turn off the H.P. cock.

(b) Ensure that the impellor has stopped turning ; wait until the fuel stops draining from the nacelle and then dry out the engine by carrying out the starting procedure with the H.P. cock in the OFF position.

(c) When the impellor has stopped turning, have the ground crew remove any surplus fuel from the jet pipe.

(d) Start the engine as in sub. paras (iv) and (v) above.

NOTE.—If for any reason the engine fails to start after two attempts the cause should be investigated before making further attempts to start.

37. **Testing the engines and services**

(i) While idling at 5,000 to 6,000 r.p.m. carry out the checks detailed in the Pilot's Check List, items 80 to 86.

(ii) To check the operation of each vacuum pump see that the artificial horizon erects properly and maintains the correct position after the vacuum change-over cock is operated.

(iii) If it is required to check the engine-driven hydraulic pump, the flaps must be lowered and raised at least four times to exhaust the accumulator, and then lowered and raised once more to check the pump.

(iv) If during engine starting a severe hammering is encountered it is due to air in the hydraulic system and should disappear when one of the hydraulic services is operated.

38. **Taxying**

(i) Carry out the checks detailed in the Pilot's Check List, items 87 and 88.

(ii) Rapid and unnecessarily frequent opening and closing of the throttles should be avoided as it will result in excessive jet pipe temperatures and possibly a momentary surge.

(iii) Response to throttle opening is slow, and it is not easy to turn without assistance from the brakes.

(iv) When taxying, fuel consumption is high, being over one gallon per minute for each engine at idling r.p.m. There is no engine-driven compressor and the brakes should be used as sparingly as possible.

39. **Take-off**

NOTE.—(a) When conditions make the use of the shortest possible take-off run essential the brakes should be applied when the aircraft is aligned on the runway and the throttles opened gradually to take-off r.p.m. The brakes should then be released.

(b) If it is necessary for any reason to check any of the engine instruments, this should be done against the brakes, prior to take-off.

(i) Carry out the checks detailed in the Pilot's Check List, items 89 to 96.

(ii) Taxy forward a few yards to straighten the nose wheel, and then open the throttles smoothly to take-off r.p.m. ; there is no tendency for the aircraft to swing.

(iii) As the aircraft accelerates to about 70 kts., ease the control column back in order to raise the nose wheel just clear of the ground. Care must be taken not to get the nose wheel too high ; this is easily done as the elevator is light and very powerful at low speeds.

(iv) The acceleration is slow and the aircraft does not unstick cleanly. At training load the aircraft should be flown off the ground at 95-100 kts. and at full load it should be pulled off at 105-110 kts.
When a ventral tank is carried the take-off run is considerably increased.

(v) When comfortably airborne, apply the brakes to stop the wheels spinning, then retract the undercarriage.

(vi) With the flaps $\frac{1}{3}$ down safety speed at full load at full take-off power is 115 kts. The aircraft should be held down, therefore, until this speed is attained. The flaps should then be raised ; no change of trim or sink results.

(vii) Unless it is necessary to clear obstacles, allow the aircraft to accelerate up to 155 kts. before commencing to climb.

40. **Climbing**

(i) The speeds for maximum rate of climb, which are also the recommended climbing speeds are given below :—

Altitude					
Sea level	...	...	...	...	...	210 knots
10,000 ft.	...	...	...	...	....	190 ,,
20,000 ft.	...	...	...	...	...	170 ,,
30,000 ft.	...	...	...	...	...	150 ,,

i.e., start to climb at 210 kts., and reduce speed by 20 kts. every 10,000 ft.

(ii) A governor on the engine-driven fuel pump restricts the r.p.m. to 16,500 for take-off, but when climbing with the throttles fully open, r.p.m. will tend to increase progressively with height. Normally the throttles should be adjusted on the climb to maintain the r.p.m. at 16,000, but to maintain a reasonable rate of climb at high altitudes, r.p.m. should be increased as necessary to 16,500 providing a jet pipe temperature of 650°C. is not exceeded.

41. Maximum range and endurance

(i) Maximum range

  (a) The recommended cruising speeds for maximum range are :—

  > 215 kts. up to 20,000 ft.
  > 200 kts. above 20,000 ft.

  (b) An alternative method, which does not appreciably reduce the range, is to cruise with the throttles set to give 15,000 r.p.m. irrespective of height or weight, the resulting I.A.S. being accepted.

(ii) Maximum endurance

  Fly at the highest altitude consistent with the prevailing weather conditions, adjusting the throttles to maintain a speed of 145 kts.

42. Position error corrections

The position error corrections at sea level are :—

From ...	120	190	above	} knots
To	190	300	300	
Add ...	2	4	6	knots

43. General flying

(i) *Stability.* Stability about all axes is satisfactory but care must be taken to ensure that aircraft which have long chord engine nacelles are ballasted correctly, otherwise the C.G. may fall behind the aft limit and longitudinal stability will then be impaired.

In bumpy conditions, specially at high speeds, the aircraft tends to " snake," i.e., to oscillate directionally.

(ii) *Changes of trim*

Operation of the undercarriage, the flaps or the air brakes promotes little change of trim. Closing the throttles at high speed, or in a dive, causes a strong nose-down change of trim.

(iii) *Controls*

(a) The trimming controls are spongy in operation, and accurate trimming therefore demands great patience, especially at high altitudes.

(b) The elevator is light and powerful but the rudder and ailerons are heavy. On aircraft which have the long chord engine nacelles the ailerons are exceedingly heavy and any rolling manœuvre is tiring.

(c) The air brakes are very effective.

(iv) *Flying at reduced airspeeds.* Use the air brakes to reduce speed to 195 knots, then lower the flaps to $\frac{1}{3}$ down, and close the air brakes. Speed may then be reduced to not less than 130 knots.

(v) *Engine handling*

(a) Rapid throttle opening may promote surging, but the engines will accelerate through the surge without damage. Surging, which is characterised by a muffled banging, can also occur on the climb at a fixed throttle setting. In these circumstances it may be overcome if the throttle setting is reduced slightly.

(b) At a fixed throttle setting the r.p.m. tend to increase as height is gained. This tendency should be checked by partially closing the throttles.

(c) Above 20,000 ft. the throttles must be handled carefully. If they are opened or closed quickly, or if r.p.m. are reduced below 10,000 combustion may cease. At very high altitudes there is a possibility of this occurring at even higher r.p.m. The fact that an engine has stopped may not be immediately apparent. It is recommended, therefore, that above 20,000 feet any reduction of speed or height should be effected by opening the air brakes.

However, should it be imperative to reduce throttle openings at these high altitudes, the engines should be throttled back in turn, in order to eliminate the possibility of both engines stopping. When it has been ascertained, by reference to the jet pipe temperature gauge, or r.p.m. indicator, that the engine is still functioning, power may be reduced on the other engine. Should an engine stop above 15,000 feet, the H.P. cock should be turned OFF immediately, and no attempt should be made to restart until height has been reduced below this altitude.

(d) Combustion may cease if negative " g " is applied for more than 15 seconds.

(e) A total of 100 gallons of fuel should be left for descending from high altitudes and for landing.

44. **Stalling**

(i) The approximate stalling speeds in knots are :—

	At training load:— 12,200 lb. (full internal fuel, no ammunition or external stores).	At 13,600 lb. (f u l l internal fuel, full ammunition and 1 × 100 gallon fixed ventral tank).	At 14,500 lb. (f u l l internal fuel, full ammunition and 1 × 180 gallon ventral drop tank).
Undercarriage and flaps up	85	85–90	90–95
Undercarriage and flaps down	75	80	80–85

On aircraft with long chord engine nacelles the stalling speeds with undercarriage and flaps down are increased by approximately 5 knots.

(ii) With the air brakes extended the stalling speeds quoted in sub. para. (i) are, in general, increased by 2-3 knots.

(iii) With the air brakes closed there is little warning of the approach of the stall except for slight tail buffeting, the onset of which can be felt some 10 knots before the stall itself. With the air brakes extended or the sliding hood open the buffeting is more pronounced and the aircraft tends to pitch as the stall is approached. Just before the stall, either wing may tend to drop ; this tendency can be checked with opposite aileron but either wing will then drop sharply and aileron snatching will be felt. If the control column is held back at the stall the aircraft will generally become inverted and recovery then involves a considerable loss in height.

On aircraft which have long chord engine nacelles the pre-stall tail buffeting is very pronounced.

(iv) Warning of the approach of the stall in a steep turn is given by aileron and elevator buffeting. Continued backward movement of the control column will then cause the inner wing to drop or the aircraft to flick out of the turn, but recovery is straightforward if the pull force on the control column is relaxed.

45. Diving

(i) Since it is inadvisable to throttle back to any extent when above 30,000 feet the air brakes must be used to avoid exceeding the limitations during a descent. Even with them open the limitations can easily be exceeded. With the air brakes open there is marked instability and considerable shuddering which must not be confused with compressibility effects.

(ii) The aircraft becomes increasingly tail heavy as speed is gained, but it can be held in the dive without trimming.

(iii) Recovery will normally be effected by relaxing the push force on the control column, but if for any reason the elevator trimming tab is used to assist in recovery it must be applied slowly and carefully since it is powerful and somewhat spongy.

(iv) Closing the throttles in a dive induces a slight nose-down change of trim ; this becomes more pronounced at high speed.

(v) The air brakes should not be closed in steep dive since this induces nose-down change of trim and, as they close, the aircraft accelerates rapidly.

46. Aerobatics

(i) The following speeds in knots are recommended :—

Roll	240
Loop	310
Half roll off·the top of the loop	310–325
Climbing roll	345

(ii) In manœuvres in the looping plane much height may be lost or gained and an ample margin must always be allowed for recovery to normal flight.

(iii) The negative " g " traps in the main tank ensure a supply of fuel for not more than 15 seconds inverted flight.

(iv) Aerobatics are prohibited when carrying a full 180 gallon ventral drop tank.

47. Approach and landing

(i) When entering the circuit carry out the checks in the Pilot's Check List, items 97 to 102.

(ii) The turn into wind should be made at approximately 130 knots, airspeed being reduced progressively so that the airfield boundary is crossed at the following speeds : At light load (all ammunition expended and approximately 50 gallons of fuel remaining) with or without a ventral tank :—

Aircraft with short chord engine nacelles	95 knots
Aircraft with long chord engine nacelles ...	100 knots

48. Mislanding and going round again

(i) A minimum of 60 gallons fuel should be allowed for going round again once.

(ii) The aircraft will climb away easily at climbing power with the undercarriage and flaps down.

(iii) Open the throttles slowly to give climbing r.p.m. and re-trim. If the throttles are opened quickly the engines may surge, causing a temporary loss of power.

(iv) Raise the undercarriage and flaps.

49. After landing

(i) Before taxying, carry out the checks in the Pilot's Check List, items 103 to 106.

(ii) To stop the engines turn off the H.P. cocks, then, when the engines have stopped, switch off the low-pressure pumps, and turn off the L.P. cocks. Then carry out the checks in the Pilot's Check List, items 110 to 119.

PART III
LIMITATIONS

50. **Engine data—Derwent Mk. 1**

The principal engine limitations using aviation kerosene, plus 1% lubricating oil are as follows :—

	R.p.m.	Jet pipe temp. °C
TAKE-OFF (5 minutes limit)	16,500 ± 100*	690
MAXIMUM CLIMBING (30 minutes limit)	16,000	650
MAXIMUM CONTINUOUS	15,400	600
OPERATIONAL NECESSITY (5 minutes limit)	16,500 ± 100*	690
IDLING AT GROUND LEVEL (10 minutes limit)	5,000-6,000	550

* This tolerance is allowed on the initial setting of the overspeed governor.

NOTE.—The jet pipe temperature limitations listed above must not be exceeded under steady conditions.

OIL PRESSURE
Normal	...	...	...	...	35 lb./sq. in.
Minimum in flight		...	...	30 lb./sq. in.
Minimum for idling		...	...	10 lb./sq. in.

OIL TEMPERATURE
| Maximum .... | ... | ... | ... | ... | 80°C. |
| Minimum for opening up... | | ... | ... | 0°C. |

51. **Flying limitations**

(i) The aircraft is designed for the duties of a single-seat fighter, but intentional spinning is not permitted. Should an unintentional spin develop, normal recovery action is effective, but considerable height may be lost before re-gaining level flight. Control forces during the spin and the recovery may be high.

31

(ii) *Maximum speeds in knots*

    (a) Diving without external stores*

Below 6,500 ft.	430
Between 6,500 ft. and 10,000 ft.	400
Between 10,000 ft. and 15,000 ft.	365
Between 15,000 ft. and 20,000 ft.	330
Between 20,000 ft. and 25,000 ft.	300
Between 25,000 ft. and 30,000 ft.	270
Between 30,000 ft. and 35,000 ft.	240
Between 35,000 ft. and 40,000 ft.	215

* When a full 180 gallon ventral drop tank is carried the aircraft is limited to 345 knots below 15,000 ft.

    (b) Lowering undercarriage ... ... 175 knots

          Undercarriage locked down ... ... 195 knots

    (c) Lowering flaps ⅓ down ... ... No limit (A relief valve in the system prevents excessive strain)

          Flaps fully down ... ... 150 knots

    (d) Opening air brakes ... ... No limit, but the brakes will not open fully above 345 knots.

(iii) *Maximum weights*

	Short nacelle aircraft	Long nacelle aircraft .
Take-off, straight flying and gentle manœuvres ...	14,750 lb.	15,300 lb.
* All forms of flying ...	13,100 lb.	13,100 lb.
Landing	12,000 lb.	12,000 lb.

* When there is any fuel in the 180-gallon ventral drop tank the aircraft is restricted to straight flying and gentle manœuvres irrespective of the all-up weight.

(iv) *Ventral drop tank jettisoning*

The 180-gallon ventral drop tank should only be jettisoned when necessary operationally. While jettisoning, the aircraft should be flown straight and level at a speed not greater than 345 knots.

PART IV
EMERGENCIES

52. **Engine failure during take-off**

With the flaps ⅓ down, undercarriage up or down, safety speed at normal load is 110 knots and with a full 180 gallon ventral drop tank is 115 knots. Providing safety speed has been attained the aircraft can be held level and the flaps can be raised without any change of trim or sink occurring. Allow the speed to build up to 130 knots, then climb away. Turn off the H.P. and L.P. cocks and the low pressure pump of the failed engine, and open the balance cock.

53. **Engine failure in flight**

(i) At low and medium altitudes the aircraft will maintain height easily on one engine at 155 knots at cruising power.

(ii) In the event of engine failure, turn off the H.P. and L.P. cocks and the low pressure pump of the failed engine, then open the balance cock (see para. 3 (iii)) but when practising single-engine flying the L.P. cock should not be turned off.

(iii) It is recommended that no attempt should be made to restart an engine above 15,000 feet (see para. 55 (i)). Practice relights should be restricted to a minimum, and should be carried out only when type A.C.2 or A.C.3 high pressure cocks are fitted.

(iv) In the event of failure of No. 1 engine it is important to conserve the batteries as much as possible. The vacuum change-over cock should be set to STARBOARD and the direction indicator synchronised with the R.I. compass while the latter is still working.

(v) If engine failure is due to an obvious mechanical defect, relighting should not be attempted. Close the appropriate L.P. and H.P. cocks and switch OFF the low pressure pump concerned. Closing the L.P. cock is a precaution against loss of fuel should the fuel pipeline be broken.

(vi) If the cause of the failure is not apparent, it is probably due to combustion ceasing either through engine mishandling or through the burner pressure dropping too low at high altitude. In this case only the H.P. cock should be turned off. This avoids starving the engine-driven fuel pump and relighting may be attempted at 15,000 ft. or below.

54. Single-engined landing

(i) Maintain a speed of at least 130 knots while manœuvring with the undercarriage and flaps up.

(ii) A single-engined landing presents no difficulty and a circuit in either direction can safely be made irrespective of which engine has failed. Lower the undercarriage and $\frac{1}{3}$ flap as on a normal circuit, maintaining a speed of 130 knots until the decision to land has been made. Then lower the flaps fully when required and use the approach speeds quoted in para. 47.

(iii) With the wheels and $\frac{1}{3}$ flap down, going round again on one engine from 130 knots using full power, involves no loss of height. If No. 2 engine has failed the undercarriage and $\frac{1}{3}$ flap should be left down for the circuit as the aircraft will climb away satisfactorily. Foot loads are heavy and although some pilots may be able to control the aircraft at speeds slightly below 130 knots it is advisable to keep at or above this speed until the decision to land is made and full flap is lowered.

(iv) If No. 2 engine has failed, the hydraulic accumulator, if fully charged, will operate :—

> Air brakes open
> Undercarriage down
> $\frac{1}{3}$ flap down
> Air brakes closed
> Flaps fully down.

The services should be operated in this sequence. If the air brakes have been operated earlier, or if for any other reason the accumulator is not fully charged there may be insufficient pressure to carry out all the above operations. If the undercarriage will not lower by accumulator pressure, it should be locked down by using the handpump. Airspeed should not exceed 160 knots, as above this speed the handpump may not lock the nosewheel in the down position. Flaps may be lowered with the handpump as desired. Even if fully charged the accumulator pressure will not be sufficient to raise the undercarriage and flaps in the event of going round again.

55. Restarting an engine in flight

(i) Attempting to relight an engine at heights above 15,000 ft. is not recommended, because although a successful relight may be achieved if it is attempted immediately flame extinction occurs, it may, if unsuccessful, jeopardise the chances of a relight below 15,000 ft.

(ii) Better re-lighting is obtained if the airspeed, and thus the r.p.m., are as low as possible when the H.P. cock is turned on.

(iii) Ensure that the H.P. cock is closed, then :—
Close the throttle.
Reduce the windmilling r.p.m. to 1,000-1,200 by decreasing the airspeed.
Press the re-lighting button and after 5 seconds turn on the H.P. cock, keeping the button pressed until the jet pipe temperature starts to rise. When the engine is running satisfactorily, with normal jet-pipe temperature, release the re-lighting button and open up to the desired r.p.m.

(iv) If an engine fails to re-start the H.P. cock must not be left on for more than 30 seconds and at least one minute should be allowed before the next attempt to start is made, to enable the engine to " dry out."

(v) If the above method fails, no attempt should be made to re-light the engine by using the normal starting method.

56. Undercarriage, flaps and air brakes emergency operation

In the event of failure of No. 2 engine, or of the hydraulic pump, the residual pressure in the accumulator should operate the services as in para. 54 (iv).
If, however, any of the services do not reach the desired setting after normal selection, lack of accumulator pressure is indicated. The handpump will operate any of the services through normal pipelines, and in this case should be used immediately.

57. Flapless landings

The initial approach should be made at about 115 knots. Very little power is required and the airfield boundary should be crossed at 100-105 knots. Speed drops off slowly and the aircraft requires a long landing run.

58. Hood jettisoning

(i) The hood may be jettisoned from inside the aircraft by pulling the handle (27) on the top right-hand side of the front panel. Before jettisoning the hood, the seat should be lowered fully and the pilot should keep his head forward and well down.

(ii) There is an external jettison handle covered by a rip patch on the port skin below the windscreen.

59. Fire-extinguishers

(i) The engine fire-extinguishers are operated electrically by two shielded pushbuttons on the top right-hand side of the front panel. They are also automatically operated by an impact switch, in the event of a crash landing.

(ii) If a fire is observed in an engine nacelle the H.P. and L.P. cocks should be turned off, the throttle closed and the low-pressure pump switched off. Airspeed should be reduced as much as possible before the fire-extinguisher is operated.

(iii) If fumes enter the cockpit the ventilating control should be set to COLD.

60. Crowbar

A light crowbar is clipped to the left-hand side of the pilot's seat.

61. Ditching

Model tests indicate that in calm weather the aircraft should ditch well.

(i) If a ventral drop tank is fitted it should be jettisoned.

(ii) The sliding hood should be jettisoned.

(iii) The safety harness should be kept tightly adjusted and locked but the R/T and oxygen leads should be disconnected.

(iv) The undercarriage should be kept retracted but the flaps should be lowered fully to reduce the touchdown speed as much as possible.

(v) The engines, if available, should be used to help make the touchdown in a tail-down attitude at as low a forward speed as possible.

(vi) Ditching should be along the swell, or into wind if the swell is not steep.